CCNB1, CCNB2, CCNA1, CCNA2, SYT1, SYT2, CKS2, CKS1B, CCNB3, SKP1, CDK1, RPS23, RPS27A, ZFAND4, RPS27, RPS27l, BUB1, BUB1B could play significant roles in the aetiology of schizophrenia by acting as points of contact between ALDH18A1 and SEC23IP (COP2).

Author: John Neville
Contact details: jwilljohnwill@gmail.com
Keywords/phrases:
Published: 18.12.2017
Addendum added: 20.12.2017
Publisher: Medresind

First edition
Copyright: John Neville

Abstract

Fourteen genes and their paralogues (CCNB1, CCNB2, CCNA1, CCNA2, SYT1, SYT2, CKS2, CKS1b, CCNB3, RPS23, RPS27A, ZFAND4, RPS27, RPS27I, BUB1, BUB1B, SKP1, CDK1), which putatively act as points of contact between ALDH18A1 and COP2 associated genes, particularly SEC23IP and CSNK1D, are identified which could play significant roles in the aetiology of schizophrenia. Many of these genes are found at the same genetic locations as deletion/duplication disorders and/or CNVs that have been reported on as being associated with schizophrenia.

Introduction

ALDH18A1 produces ornithine from glutamate and, via PYCR1, proline. Low proline and low ornithine are diagnostic indicators for ALDH18A1 disorders. PYCR1 performs the opposing function to PRODH.

ALDH18A1 disorders include cutis laxa and spastic paraplegia. PYCR1 disorders include cutis laxa and Geroderma Osteodysplasticum. PRODH is part of the Di George 22q11.21 deletion region which is functional for schizophrenia and has been associated with schizophrenia.

A proband in a case of familial psychosis has low ornithine and low proline. He and his family suffer from a pattern of ALDH18A1 disorder symptoms and PYCR1 disorder symptoms. As such, there is excellent evidence that there is probably abnormal function of ALDH18A1 in the familial disorder.

There is also good evidence that there could be COP2 transport disfunction in the familial disorder. SAR1B is perhaps the key protein involved in COP2 transport. SAR1B is located at 5q31.1.

The proband suffers symptoms of SAR1B Chylomicron Retention Disorder. The proband's diarrhea is responsive to vitamins and fatty acid supplementation, which is consistent with treatment responsiveness to fatty acids and specific vitamins which are used to treat diarrhea in Chylomicron Retention Disorder. The proband carries multiple SNPs at 5q31.1 that are associated with or significantly associated with schizophrenia in Irish high-density schizophrenia families.

It was previously suggested that the proband's family suffered from very mild symptoms of Smith Magenis Disorder which results from defects at 17p11.2. Particular reference was subsequently made to the possible significance of SREBP1 (part of the Smith Magenis deletion region) function in the proband's familial disorder because of the significant role that SREBP1 plays in SAR1B mediated COP2 transport. It was suggested that altered SREBP1 function could account for the symptoms of Smith Magenis disorder in the proband's family.

Additionally, it was noted that Parkinson's disease or symptoms of Parkinson's disease (often very mild) were present in family members who had suffered symptoms of psychosis, and it was pointed out that the proband's aunt had also suffered from treatment responsive rheumatoid arthritis.[1]

In this paper various patterns are identified amongst key genetic locations which have previously been identified as being possibly important in the proband's familial disorder.

Significantly, it is pointed out that genes that putatively connect ALDH18A1 to COP2 transport associated genes (particularly SEC23IP and CSNK1D) are found at genetic locations which correlate with the genetic locations of deletion/duplication disorders and CNVs that are reported on in schizophrenia. It is concluded that fourteen connecting genes and their paralogues (Ccnb1, Ccnb2, Ccna1, Ccna2, Syt1, Syt2, Cks2, cks1b, Ccnb3, Rps23, Rps27A, Zfand4, Rps27, Rps27l, Bub1,

Bub1B) could play significant roles in the aetiology of schizophrenia. Also, it is pointed out that any further genes which are points of contact between ALDH18A1 and COP2 associated genes could be significant in the aetiology of schizophrenia.

Finally, it is hypothesized that schizophrenia associated CNVs which impact on COP2 function could result in the selection of compensatory SNPs that also act on COP2 function. It could be that the compensatory SNPs, and not the CNVs, are causative of the symptoms of schizophrenia.

Materials

5q31.1, 5q31.2, 22q11.21, 3q22.3 and 17p11.2 as core locations in the proband's familial disorder

The following is a list of genes located at 5q31.1 (the location of SAR1B), 22q11.21 (the PRODH and Di George deletion location), 5q31.2, 3q22.3 and 17p11.2 (the Smith Magenis deletion location) where schizophrenia associated SNPs are carried by the proband. 3q22.3 is included here because the proband carries schizophrenia associated SNPs at several different genes at this location.

Stag1
Acsl6
IL3
Comt

Tgfbi

Smad5

Spry4

Neurog1

PRODH

Slc25a48

Zdhhc8

Pi4ka

Etf1

Srebp1

17p13.1. 22q11.21. 17q25.3. 5q31.1.

It was previously proposed that in addition to 5q31.1 and 22q11.21, 17p13.1 and 17q25.3 could be core locations in the proband's familial disorder. 17p13.1 is the location of TP53 which regulates ALDH18A1 function and 17q25.3 is the location of PYCR1.

It was also previously noted that there was evidence of runs of homozygosity at genetic locations that were on or which regulated the pathways thought to be over and/or under-functioning in the proband's disorder, including at genes located at 17p13.1, 22q11.21, 17q25.3, 5q31.1, and that some of these locations corresponded with locations where runs of homozygosity had been reported on in the literature on schizophrenia.

For the purpose of this paper, it should be stated here that in the proband, there is evidence of runs of homozygosity at the following genes located at 17p13.1, 22q11.21, 17q25.3, or 5q31.1:

TP53
PRODH
PYCR1
SAR1B
IL3
Acsl6
Tgfbi
Smad5
Slc22a4

Other genes and locations where the proband carries SNPs associated with schizophrenia, Parkinson's disease or rheumatoid arthritis.

6p22.3. 6q23.2. 20p12.2. 22q13.1. 1p36.22. 2q31.1. 6p21.32. 12q12. 2q32.1 5p12. 15q26.1. 14q32.2. 3p26.3/.2. 4q33. 10q12.33. 11q12.1. 8q24.3. 15q24.3 4q22.1. 10q24.32. 6p22.1. 18q22.2. 6p21.33 12q24.12. 8p21.3. 2q14.1. 1q32.1. 4q27

Kiaa1189
Adad1
Lingo1
Tsnare1
Tmx2-ctnnd1

Cacnb2

TAar6j

Dtnbp1

Snap25

Cacna1i

Csf2rb

MTHFR

GAD1

Hladrb1

Hladrb6

Lrrk2

Znf804a

Hcn1

Furin

bcl11b

Cntn4

Clcn3

Slc18a1

Csf2ra

COP2

The following are a list of some COP2 transport associated genes which are located at the core genetic locations (17p13.1. 22q11.21. 17q25.3. 5q31.1, 5q31.2, 3q22.3) listed above or at 10q24.1 (the location of ALDH18A1. A small minority of those listed are found at locations very close to these core locations.

SAR1B

SEC24A

Csnk1d

Csnk1a1

Sec31b

Trappc1

SREBF1

Ulk2

Rptor

MTOR

Nup85

Pemt

Pld2

PER1

BECN1(17q21.3)

PI3K

Previously it was suggested that in the proband's familial disorder, there may be a failure or partial failure of the PI3K product PI3P and/or PI4P binding to the phospholipase A1 SEC23IP as part of the COP2 transport process. The following are a list of some PI3K connected genes which are located at the core locations listed above or at 10q24.1 (the location of ALDH18A1).

PI4KA

Pik3cb

Pik3r4

Pik3ap1

Pi4ka

YWHAE

PIK3CD (1p36.22 which is a location where the proband carries one or more SNPs associated with schizophrenia/rheumatoid arthritis/Parkinson's)

Genes that connect ALDH18A1 associated genes to COP2 transport associated genes.

If SEC23IP, CSNK1D (a casein kinase located at 17q25.3, the same location as PYCR1, which phosphorylates SEC23p) and ALDH18A1 are inserted into the STRING functional protein network, and STRING is asked to make more connections (medium confidence and excluding text mining), STRING eventually connects ALDH18A1 to SEC23IP and CSNK1D via CKS2, CCNB1, CKS1B, SKP1 and CDK1. If various combinations of these genes and the genes listed above are inserted into STRING, it makes at least 9 other connections between ALDH18A1 and COP2 transport associated genes, making a total of 14 connecting genes.

A pattern is immediately noticeable in relation to the repeating genetic locations amongst these 14 putatively connecting genes and the paralogues of these 14 genes which are listed below. There is a clear overlapping between the genetic locations of some of these fourteen genes and the paralogues.

Ccnb2 at 15q22.2 paralogue is ccnb1 at 5q13.2

Ccnb1 at 5q13.2 paralogue is ccnb2 at 15q22.2

Ccna2 at 4q27 paralogue is ccna1 at 13q13.3!

Ccna1 at 13q13.3 paralogue is ccna2 4q27

Syt1 at 12q21.2 paralogue is syt2 at 1q32.1

Cks2 at 9q22.2 the paralogue of which is cks1b which is also at 1q21.3

Cks1b at 1q21.3 paralogue is cks2

CCNB3 at xp11.22 paralogue is ccna2 at 4q27

RPS23 at 5q14.2

SKP1 at 5q31.1 – the location where the proband carries multiple SNPs at multiple genes that are significantly associated or associated with schizophrenia.

CDK1 at 10q21.2

RPS27A 2p16.1 the paralogue is zfand4 at 10q11.22

RPS27 1q21.3 paralogue is RPS27l at 15q22.2

BUB1 at 2q13 paralogues is BUB1B at 15q15.1

It is then noticeable that the genetic locations of many of these 14 genes and their paralogues correspond with the genetic locations of deletion/duplication disorders and/or CNVs that are reported on is schizophrenia, or in some cases with schizophrenia associated locations.

2p16.1 2p15 to 2p16.1 microdeletion

2p16.1. Same location as fbln3

2p16.1 variants associated with schizophrenia

2p16.3. Microdeletion schizophrenia

2p16.3 associated with rheumatoid arthritis and Parkinson's

1q21.3. Microdeletion

1q21.1 - 21.2 microdeletion schizophrenia

1q21.1 associated with spastic paraplegia

15q22.2 microdeletion

15q22.2 location of PARK23 causative of Parkinson's

15q13.2 cnv schizophrenia

15q13.3 microdeletion schizophrenia

15q11.2-13 duplication schizophrenia

5q13.2 microdeletion

5q11.2 to 13.3 duplication schizophrenia

4q27 to 28.1 microdeletion

4q deletion with schizoaffective Disorder

12q21.2 deletions

12q21.2 CNVs Parkinson's

12q deletions including psychiatric

9q22.2 microdeletion

9q21.13 cnv bd

10q11.22/q11.23 deletion

10q11.21 to 22. Cnv schizophrenia but more bd

13q13.3 to q21.3 deletion

13q13.3 duplication schizophrenia

1q32.1 microdeletion

1q32.1 evidence of linkage schizophrenia and 1q32 = linkage schizophrenia

Xp11.22 to xp11.23 microduplication

Xp11 significantly associated with schizophrenia

5q14.2 associated with retinues pigmentosa a ALDH18A1 symptom

5q14.3 microdeletion schizophrenia

5q14.3 deletion disorder

2q13 deletion disorder
2q13 location of fbln7
2q13 duplication schizophrenia
10q21.2 a location associated with schizophrenia

Discussion

There are patterns in the core genetic locations mentioned above which link these genetic locations to genes that play central roles in COP2 transport, as well as to genes that play central roles in PI3K function.

Additionally, there is a pattern involving the 14 genes mentioned above that putatively connect ALDH18A1 to SEC23IP/CSNK1D function and/or COP2 transport, which links the genetic locations of some of these genes to the genetic locations of some of their paralogues.

Finally, there is a pattern which demonstrates that the genetic locations of many of these 14 genes and their paralogues play central roles in the aetiology of schizophrenia.

All of these patterns still require explanation.

However, evidence is accumulating as to the significant role SEC23IP binding probably plays in the proband's familial disorder and potentially in schizophrenia more generally. Clearly the 14 genes identified above and their paralogues could themselves play significant roles in the aetiology of schizophrenia.

Also, it is evident that any gene which is at the point of contact between ALDH18A1 and COP2 associated genes, particularly SEC23IP and CSNK1D, could be significant in the aetiology of schizophrenia.

It is hypothesized that schizophrenia associated CNVs which impact on COP2 function, whether in relation to transport, autophagy or both, could result in the selection of compensatory SNPs that also act on COP2 function. The patterns identified in this paper may indicate that the selection of compensatory SNPs could result from purpose-driven rapid selection, which was previously proposed by the writer. However, the natural selection of these compensatory SNPs cannot yet be conclusively ruled out. It could also be the case that the compensatory SNPs, and not the CNVs, are causative of the symptoms of schizophrenia.

Addendum added 20.12.2017

It was previously hypothesized that a failure or partial failure of ALDH18A1/PYCR1 to produce the specific proline residues required for SEC23IP binding to PI3P could be causative of symptoms in ALDH18A1 spastic paraplegia and also of symptoms in ALDH18A1/PYCR1 cutis laxa, and that if so, any such missing proline residues may be viable treatment targets.[2]

Given that PRODH performs the opposing function to PYCR1, then the same possibility may apply in relation to 22q11.2

Parkinson's disease as well as 22q11.2 schizophrenia. Defects at 22q11.2 may result in a reduced production of the proline residues required for the binding of SEC23IP, thus partially interrupting COP2 transport and increasing COP2 autophagy. This partial or complete failure in SEC23IP binding might be causative of symptoms of 22q11.2 Parkinson's disease and/or schizophrenia. As such, supplementing the appropriate proline residues may be a highly effective treatment in 22q11.2 Parkinson's disease and/or schizophrenia.

Here it is recalled that ALDH18A1 defects result in spastic paraplegia, and that SEC23IP is a phospholipase A1 that contains a DDHD region. DDHD1 and DDHD2 are also phospholipase A1s and defects at both DDHD1 and DDHD2 are also causative of spastic paraplegia.

The connections between 22q11.2, SEC23IP and schizophrenia are detailed above. In relation to Parkinson's disease, it is noted here that if a STRING search is carried out which lists SEC23IP, ALDH18A1, CSNK1D as well as CKS2, CCNB1, CKS1B, SKP1 and CDK1, and the STRING program is asked to make connections, it produces links to CUL1 and FBXW7, which are associated with Parkinson's disease. It is reported that SREBF1 and FBXW7 both have a conserved function in promoting mitochondrial translocation of Parkin and subsequent mitophagy.[3] A failure of SEC23IP binding being causative of symptoms of Parkinson's is made a particularly interesting possibility because of the role of SREBP1 also plays in COP2 transport and the role it appears to play in the proband's familial disorder. Additionally, some of the fourteen genes mentioned above and their paralogues have connections to Parkinson's disease which are noted above.

Declaration

The writer has no formal medical training and is not a professional researcher.

Conflict of interest

Competing interest - none declared.
The writer is the proband.

References:

[1] Neville J. Describing for the first time a COP2 transport defect that causes psychosis. 08.02.2017. Medresind.

Neville J. Purpose-driven rapid evolutionary selection: a hypothesis that proposes that rapid mutational changes may occur that support a pathway that is over-functioning and/or under-functioning and that this over-function and/or under-function may result from an immune response reaction and/or overuse of a body part or organ. 15.01.2016. Medresind.

Neville J. A hypothesis that proposes that immune response reactions may cause the rapid acquirement of pathologically significant SNPs, allele combinations and/or runs of homozygosity in rheumatoid arthritis and Parkinson's disease. 18.10.2015. Medresind.

Neville J. A hypothesis that proposes that many SNPs that are markers of schizophrenia are required for immune response and that these SNPs may be rapidly accumulated as part of one or more immune response reactions. 06.09.2015. Medresind.

Neville J. The successful treatment of a patient with psychosis who carries SNPs that are significant markers of schizophrenia in Irish high density schizophrenia families and who has MTHFR deficiency. 23.06.2015. Medresind.

Neville J. Analysing and attempting to connect the genetic and metabolic derangements underpinning a disorder which is linked to schizophrenia in Irish high density schizophrenia families. 22.03.2015. Medresind.

[2] Neville J. A hypothesis that proposes that in ALDH18A1 spastic paraplegia, the failure to produce specific proline residues required for the binding of SEC23IP as part of the COP2 transport process could be causative of symptoms and may be a viable treatment target. 05.10.2017. Publisher: Medresind

[3] Ivatt RM, Sanchez-Martinez A, Godena VK, Brown S[2], Ziviani E, Whitworth AJ. Genome-wide RNAi screen identifies the Parkinson disease GWAS risk locus SREBF1 as a regulator of mitophagy. Proc Natl Acad Sci U S A. 2014 Jun 10;111(23):8494-9. doi: 10.1073/pnas.1321207111. Epub 2014 May 27.